EXPERIMENT & THEORY
IN PHYSICS

CAMBRIDGE
UNIVERSITY PRESS

University Printing House, Cambridge CB2 8BS, United Kingdom

Published in the United States of America by Cambridge University Press, New York

Cambridge University Press is part of the University of Cambridge.

It furthers the University's mission by disseminating knowledge in the pursuit of education, learning and research at the highest international levels of excellence.

www.cambridge.org
Information on this title: www.cambridge.org/9781107665668

© Cambridge University Press 1943

First published 1943
First paperback edition 2014

A catalogue record for this publication is available from the British Library

ISBN 978-1-107-66566-8 Paperback

Experiment and Theory in Physics

BY

MAX BORN

M.A., Ph.D., Sc.D.h.c., F.R.S.

TAIT PROFESSOR OF NATURAL PHILOSOPHY
UNIVERSITY OF EDINBURGH

Cambridge

At the University Press

1943

NOTE

This paper represents in a slightly expanded form an address given to the Durham Philosophical Society and the Pure Science Society, King's College, at Newcastle-upon-Tyne on 21 May 1943

EXPERIMENT & THEORY
IN PHYSICS

<hr>

IT is natural that a man should consider the work of his
hands or his brain to be useful and important. Therefore nobody will object to an ardent experimentalist
boasting of his measurements and rather looking down
on the 'paper and ink' physics of his theoretical friend,
who on his part is proud of his lofty ideas and despises
the dirty fingers of the other. But in recent years this
kind of friendly rivalry has changed into something
more serious. In Germany a school of extreme experi-
mentalists, led by Lenard and Stark, has gone so far
as to reject theory altogether as an invention of the
Jews and to declare experiment to be the only genuine
'aryan' method of science. There is also a movement in
the opposite direction which—though not racial—is not
much less radical, claiming that to the mind well
trained in mathematics and epistemology the laws of
Nature are manifest without appeal to experiment. Two
distinguished astronomers, Milne and Eddington, follow
this philosophy, though it seems to lead them in rather
different directions.

It is not my purpose here to discuss any of the fascinating theories of these men; but I wish to direct your
attention to Eddington's philosophy, which proclaims
the triumph of theory over experiment. I am a theoretical physicist (of Jewish origin) and might be expected to rejoice in this philosophy. But I do not
rejoice.

On the contrary, I consider these ideas to be a considerable danger to the sound development of science. It is this conviction which has induced me to accept your suggestion of this difficult subject. However, I do not wish to argue with Eddington on deep philosophical principles, nor to compete with him in his unsurpassed mastery of dialectics in controversy. What I wish to show you in a simple way is the mutual relationship between theory and experiment in the actual historical development of science, and to offer a balanced opinion on the present situation and future possibilities.

But even this modest programme is not easy, because of the fact that an active scientist has little time to spend on the history of science. I have read very little, far too little, of the original literature, and the greater part of my knowledge is second-hand, taken from textbooks, handbooks and encyclopaedias. There are, however, two encouraging points. I know a few of the great classical masterpieces of mathematics and physics, enough to be certain about the historical and personal background on which they have grown. And secondly, I am old enough to say that in my own lifetime I have watched the development of modern physics, which means a very considerable part of the whole of physics. It seems to me that this provides sufficient material to form an opinion.

Scanning the history of science we notice a kind of cycle, periods of experimental expansion alternating with periods of theoretical development. Theories have a tendency to become more and more abstract and general. They culminate in principles which are first opposed by the philosophers, but later assimilated. As soon as they have become a part of a philosophical system there begins a process of dogmatisation and petrification. This feature is already noticeable in the oldest quantitative sciences, mathematics and astronomy. There is no doubt that the first geometrical knowledge discovered by the Sumerians, Babylonians and Egyptians was purely empirical. The Greeks discovered the logical interdependence of geometrical facts and founded the first deductive science as formulated in Euclid's work. If you are a modern mathematician you can of course look at geometry as a product of pure thinking, taking the axioms and postulates as definitions and the whole system as an entertaining game. But that is certainly not what the Greek philosophers meant their geometry to be: they believed they were dealing with properties of real things. The fact that the predictions of their theories were confirmed by experience in all cases led to the conviction that the axioms of Euclidean geometry contain final truth.

The Euclidean system has lived 2000 years. It has survived the decline and fall of the Graeco-Roman civilisation, and all the later upheavals of history. It went through all phases of more or less conscious dogmatisation. Even after the dawn of the modern scientific age with its critical revision of traditional opinions the actual validity of Euclid's statements was not doubted, but its possibility was made the object of philosophical speculations. Kant took it for granted that we have some direct and exact knowledge about certain things—space, time, causality, etc.—and explained it by the assumption that actually we have to do not with the things themselves, but with the forms of our intuition of these things. It is plausible that these forms of thinking are given to us *a priori*, that is prior to experience. Kant's main example of *a priori* knowledge were the theorems of geometry, *ipso verbo* understood to mean Euclid's canon.

The idea that we can produce knowledge *a priori* has its roots in the historical fact of the persistence of Greek geometry, which was replaced by a more general theory only in our own time. The real reason for the longevity of Greek geometry is the accuracy with which it describes the behaviour of bodies in our terrestrial surroundings. The first doubts were raised not on account of experimental evidence, but on logical grounds. Some mathematicians found one of Euclid's axioms, that about parallel lines, less evident than the others and began to wonder whether it could not be proved from the rest. All efforts to do this were in vain, and in the end the attempt was made (first by Gauss, but not published; then independently by Bolyai and by Lobachevsky) to prove the independence of the axiom of parallels by constructing a system of geometry in which it did not hold. These constructions of non-Euclidean geometry were successful. Gauss even made measurements in order to find out which geometry is valid in the real world. He and his successor Riemann clearly realised the empirical character of geometry. Riemann created the mathematical foundations on which Einstein, in our own time, succeeded in reducing geometry to a part of physics by his general theory of relativity.

The history of astronomy is parallel to that of geometry with the difference that some of the Greek philosophers already had clear ideas about the spherical

4

shape of the earth, the central position of the sun in the planetary system and about the real distances between its members, ideas which were lost or suppressed in the dark ages. The Church had accepted Greek philosophy and science in the form given to it by Aristotle and Ptolemy. Looking at this historic phenomenon from our point of view we may say that the stagnation of science in the middle ages is due to an excessive veneration for the mind as opposed to material phenomena, leading to a preference for theoretical speculation rather than experiment.

Indeed, the beginning of modern science in the Renaissance consisted in a new philosophy, which considered systematic experiment to be the main source of knowledge. Francis Bacon was its prophet, Galileo and Newton its real founders. Scholastic philosophy was already assailed by Descartes and other philosophers, who used mainly logical and metaphysical arguments; the theories of the Universe of these rationalists, however, seem to us unconvincing, as they are not based on sufficient evidence of observation or experiment. For the essential distinction between our time and the middle ages consists in the renunciation of tradition and the establishment of experience as the true source of knowledge. The Renaissance meant not only the rediscovery of Greek literature, but a revival of the Greek spirit, of the sceptical and at the same time constructive attitude of Greek philosophy. Then the method of inductive reasoning was established, which leads from single observations to general laws. This method itself can be made an object of philosophical analysis; it is clear that it presupposes not only a fundamental belief in the existence of natural laws, but also criteria for distinguishing genuine regularities from accidental ones, and other principles of this kind. But I cannot dwell on these abstract problems. I only wish to state that the revolution which replaced scholasticism by modern science has dethroned the deductive method from its dominating position and put it in its proper place. Both Galileo and Newton were absolutely clear about the inductive character of the new philosophy; the theories which they formed by synthesis of experimental results were used for suggesting new experiments, and if these tests were favourable the theory was considered as confirmed. That is the legitimate method of science, a blending of deduction and induction, which is described in innumerable textbooks. But it is not the whole story.

Galileo and Newton were both anxious to avoid metaphysical speculation (*hypotheses non fingo*). But a short time later, when the laws of mechanics were fully known, we find attempts to derive them from principles which by their formulation suggest some non-empirical origin. The most successful of these principles is that of least action. Maupertuis was certainly led to it by a teleological idea; Nature was supposed to act like a human being, with a definite purpose which it tries to attain with the smallest amount of 'action' possible. Why the mathematical expression which he gave for this action should be so dear to Nature as to be spent parsimoniously is of course difficult to explain. We know to-day that the actual motions do not correspond to real minima of action except for short time intervals, but to stationary states, and we consider the principle of least action only as a very useful and powerful tool for condensing complicated differential equations in a short expression.

The power of this principle, in the form given to it by Hamilton, is seen by the fact that not only classical mechanics of particles and rigid bodies, but also elasticity and hydrodynamics, electromagnetism and all the modern field theories connected with ultimate particles (electron, proton, meson) can be formulated with its help. To give an example, let us consider electromagnetism.

For this purpose assume the existence of a scalar potential Φ and a vector potential A, and introduce for the sake of abbreviation the vectors

$$E = -\operatorname{grad} \Phi - \frac{1}{c}\frac{\partial A}{\partial t}, \quad H = \operatorname{curl} A. \tag{1}$$

Then the principle of action for electrodynamics in empty space is given by

$$\delta \int\int \tfrac{1}{2}(E^2 - H^2) \, dv \, dt = 0, \tag{2}$$

where dv is the element of volume and the integration is extended over the space and time considered, while the symbol δ means a small variation of the potentials.

The results of this variation are conditions in the form of differential equations, and these turn out to be Maxwell's equations

$$\operatorname{curl} H = \frac{1}{c}\frac{\partial E}{\partial t}, \quad \operatorname{curl} E = -\frac{1}{c}\frac{\partial H}{\partial t} \tag{3}$$

for empty space, provided E and H are interpreted as the vectors of the electric and magnetic field.

7

The variational principle has something convincing in the way it condenses a great domain of phenomena in one short expression, and this perfection is further enhanced if it is considered with the eye of the mathematician, who has learnt the principle of relativity and knows that E and H together form a so-called six-vector having definite transformation properties for changes of the frame of reference, i.e. Lorentz transformations of space and time. For there exist only two invariants, $E^2 - H^2$ and $(E.H)^2$, and as the electrodynamic action must be invariant, it can be only a function of these; add to this the postulate that the resulting equations ought to be linear, then the action must be quadratic, and you are led directly to the expression given above.

This seems to be straightforward reasoning from first principles. Given the knowledge and the penetrating brain of our mathematician, Maxwell's equations are a result of pure thinking and the toil of experimenters antiquated and superfluous.

I need hardly explain to you the fallacy of this standpoint. It lies in the fact that none of the notions used by the mathematicians, such as potential, vector potential, field vectors, Lorentz transformations, quite apart from the principle of action itself, are evident or given *a priori*. Even if an extremely gifted mathematician had constructed them to describe the properties of a possible world, neither he nor anybody else would have had the slightest idea how to apply them to the real world. The problem of physics is how the actual phenomena, as observed with the help of our sense organs aided by instruments, can be reduced to simple notions which are suited for precise measurement and used for the formulation of quantitative laws. It was a long way from the first observation of simple electric phenomena, like the attraction of small particles or the observation of small sparks, to the concept of electric field and potential, a still longer way to the interaction of these with the corresponding magnetic forces, and to the system of Maxwell's equations connecting them.

When I was a student, forty years ago, the idea of the field *in vacuo* was extremely strange to us, assimilable only with difficulty. From this point to the full development of relativity with its formal apparatus of Lorentz transformations, its invariants, covariant vectors and tensors is again a long and tedious journey. The relativisation of time was forced upon us: Einstein's paper was later than the experiments of Michelson and Morley, and even Lorentz himself was reluctant to give up his absolute stationary ether and to accept the equivalence of different times admitted by his transformations.

The order of historical events clearly shows the true position of the variational principle: It stands at the end of a long chain of reasoning as a satisfactory and beautiful condensation of the results. It may even have helped to find these results (though I doubt it in this case of electromagnetism). But it has little to do with the formation of the fundamental new concepts which are the characteristic feature of electrodynamics. The revolutionary conception which distinguishes electrodynamics from classical mechanics is that of the field. One can see in Faraday's work how it sprang from his observations of dielectric, paramagnetic and diamagnetic properties; but it needed Maxwell's powerful mathematics to formulate it. However, that was not mathematics pure and simple, but an amazing act of divination. The facts known at the time would have led (for the vacuum) not to the complete set of equations (3), but instead of the first one to

$$\operatorname{curl} H = 0.$$

Maxwell's decisive step consisted in adding the missing term $\frac{1}{c}\frac{\partial E}{\partial t}$ without proper empirical foundation, first guided by mechanical models of the ether, later by reasons of mathematical perfection or beauty, or however you may describe the act of genius. It is this term which leads to the prediction of waves with the finite velocity c, to the electromagnetic theory of light, to wireless and all that modern radio-engineering stands for.

Indeed this is a brilliant example of the possibilities which exist for the theoretical physicist: he can trace deficiencies in the perfection of a theory and can try to amend them by what you may call 'mathematical guessing'. If he is successful, if the modified theory predicts phenomena confirmed by new experiments, we may call these 'synthetic' predictions. This kind of prophecy is rarer but much more impressive—at least in my opinion—and generally of much greater importance than the normal 'analytical' type of scientific prophecy based on well-established theory. Of the latter kind there are so many examples that it is difficult to pick out a characteristic one. They happen in the everyday life of a physicist or engineer who designs an apparatus and expects it will work 'according to plan'.

If you enter a room and you see the head and body

of a man behind a desk you are sure that he will have legs though you do not see them, and you are not astonished if it turns out that you have been right with this prediction (alas, you may be wrong, he may be a cripple). Well, that is exactly what happens in the ordinary course of science, with the only difference that the shapes, not observed but to be expected, are less obvious than that of the human body and have to be deduced by some calculation.

On the other hand what I mean by a synthetic prediction can be illustrated by the remote side of the moon, which is invisible like the legs of the man behind the desk. Let us assume that, at the time when the spherical shape of the visible part of the moon was discovered, there was a school of philosophers who insisted that you have always to make the simplest assumption and that a plane is the simplest surface. Then the astronomers would have declared that the moon is a hemisphere with a plane back, in perfect agreement with the then available observations. However, under the influence of other experiences, the ideas of what is simple change. One day a man found that the circular edge where sphere and plane meet was against his idea of simplicity, and that a complete sphere was a more perfect surface. I shall not report to you the protracted and violent discussions about the philosophical idea of perfection and simplicity which ensued and led, at the time of the Surfaceist Government, to the trial of the modernist before the inquisition and his burning at the stake. But some time later new observations disclosed the existence of little oscillations of the moon which make small parts of the invisible side visible and show the absence of the edge, the continuity of the surface. Now, the new theory was generally accepted, and its author celebrated as a martyr of truth; and when the moderate Moon-sheviks were overcome by the radical Ball-sheviks, it became dangerous to deny that the moon was a perfect ball—although still nobody had seen the more central parts of the remote side.

Maxwell's addition of the missing term is just such a smoothing out of a roughness of a shape, though this shape is here a mathematical structure of a more refined type than a sphere.

Allow me to enlarge a little on this conception of shape which I have used. What I mean is that idea which modern psychologists (von Ehrenfels, Koehler, Wertheimer and others) have introduced under the German term 'Gestalt' in order to describe the elementary processes connected with the perception of sense impressions. The experimental fact is that simultaneous sense impressions are not independent of one another, like a mosaic, but form a psychic unit of a definite shape. You do not see coloured specks and compose them by a secondary process, but you see the man behind the desk. A good example of a shape is a melody; it is heard together with the component tones, but is obviously more than these component tones. It is a favourite idea of mine that the question 'what is the reality described by the formulae of theoretical physics?' can be answered by applying the same conception with the supplementary remark that the 'shapes' of physical things are the invariants of the equations. These have the same kind of reality—I mean: objective reality in the external world, as any shape of familiar things, for instance that of the human body. And I cannot see that analytic prediction in science differs much from the everyday procedure without which we could not live, whereby we expect that a shape recognised by a few criteria is complete with all other properties characteristic for it. A synthetic prediction, however, is based on the hypothetical statement that the real shape of a partly known phenomenon differs from what it appears to be. If confirmed by experiment it produces new knowledge, and although hypothetical it is a legitimate method. But its success depends in a high degree on intuition and can hardly be learnt.

I think that this distinction is useful in order to appreciate the value of scientific discoveries. Let us illustrate it by a few examples.

One of the most celebrated cases is the discovery of the planet Neptune by Galle according to the theoretical predictions made by Adams, and independently by Leverrier, on the basis of small perturbations in the motion of other planets. This was an admirable feat of mathematical skill and endurance, and of confidence in the results. But without minimising these one can say that it did not widen the scope of theory; it was an analytical prediction, an application of the well-known shapes of Newtonian mechanics.

Very similar is the situation with respect to the celebrated prediction of conical refraction by Hamilton, which is often quoted as an example of the power of theory. I do not in the least deny this power. But there is no doubt that this discovery was based on a given theory and consisted in the skilful unravelling of the involved properties of Fresnel's wave surface. From the standpoint of the classification suggested it belongs to the analytical class.

It is quite different with Einstein's prediction of the deflection of light by the sun derived from his theory of general relativity. For this theory is a fundamental generalisation of Newton's theory. I wish to use this example to show you that if Einstein guessed he had a very solid basis for his guessing in physical facts, so that it is justifiable to use the word 'synthesis' for it. The act of scientific imagination consists in divining the importance of a fact which in this case was known from Newton's own time, but had failed to excite the curiosity of many generations of physicists. This fact is the proportionality between mass measured by inertia and mass measured by gravity which was assumed by Newton and later confirmed by Bessel, Eötvös and others with an extremely high degree of accuracy. The problem of explaining this was beyond the reach of Newton and his successors; but the strange thing is that it did not strike these successors for two centuries as a problem at all. The possibility of a solution depended on a long sequence of research: the replacement of the idea of force acting at a distance by the conception of a field spreading with finite velocity; the establishment of linear relativity, and last but not least the hypothetical generalisations of Euclidean geometry by Gauss, Riemann, Ricci, Levi-Cività and others. All this was needed, and it was focused in Einstein's mind by the riddle of the two aspects of mass. The new theory is a gigantic synthesis of a long chain of empirical results, not a spontaneous brain wave. General relativity expresses physical laws in a geometrical language and at the same time makes geometry into a part of physics. It has an *a priori* appearance in a similar or even higher degree than Euclid's geometry. This is due to its mathematical perfection, without which we would not acknowledge it as a theory at all. But if we feel so satisfied that we consider it as final we are wrong. It soon turned out that general relativity was not final. It did not help in understanding the nature of matter, the existence of different ultimate particles and fields. Generalisations have been attempted by Einstein himself, by Weyl, Eddington, and others. But the chance of a correct guess seems to be small. So far nothing definite has come out of it apart from the fact that there is ample scope for possible theories beyond Einstein's original model.

Let us now consider the other fundamental branch of modern physics, quantum theory. It was preceded by a series of experimental discoveries—cathode rays, X-rays, radioactivity, etc.—many of which have later turned out to be beyond the power of classical mechanics and electrodynamics. But none of them led directly to the discovery of the quantum of energy; you know that Planck (in 1900) was driven to it—I should say 'in despair'—by the failure of the classical laws to account for the properties of radiant heat. He discovered his radiation formula by an interpolation of semi-empirical laws for very long and very short waves, and afterwards tentatively suggested his interpretation in terms of finite quanta of energy. The long period of twenty-five years before the appearance of quantum mechanics is characterised by the accumulation of more and more empirical evidence for the reality of this quantum and of the complete inadequacy of the classical concepts to deal with it. I can mention only some outstanding discoveries: Einstein's explanation of the photo-electric effect; his theory of specific heat of solids; Bohr's interpretation of Ritz's combination principle in line spectra and its experimental verification by Franck and Hertz, followed by Bohr's general theory of atomic structure and of the periodic table of elements; the Compton and the Stern-Gerlach effects; the disentanglement of band spectra and the discovery of quantum rules in them; and innumerable others. The vague ideas about the quantum became clearer with each new discovery; small modifications of the fundamental principles would not do, but a thorough revolution was in the making.

Quantum mechanics which was the result of this process has two apparently independent roots, matrix mechanics and its generalisations (Heisenberg, Born, Jordan, Dirac) and wave mechanics (de Broglie, Schroedinger). Before discussing these ideas from the standpoint of physics let me say a few words about the mathematical tools. Both aspects of quantum mechanics are in a large degree based on the work of Hamilton. This is often acknowledged for wave mechanics; Hamilton prepared the way by disentangling the relation between geometrical optics and wave theory, and by demonstrating the close analogy between Fermat's principle in optics and Hamilton's own formulation of the principle of least action in dynamics. But the other form of quantum mechanics, which is characterised by the use of matrices and operators, must also be traced to a fundamental conception of Hamilton's. We celebrate this year the centenary of Hamilton's invention of the quaternions, the first example of non-commuting algebra. This colourless expression does not stir the imagination as does the phrase non-Euclidean geometry, which clearly indicates the break with an ancient tradition of thought and the dawn of a new epoch. But Hamilton's work marks a turning-point of the same order of magnitude. He tried to generalise the presentation of vectors in a plane by complex numbers $z = x + iy$ to vectors in space. The idea was to have an analytical counterpart to geometrical constructions (for instance: the 'geometrical sum' of two vectors a, b is represented by the analytical process $a + b$).

Hamilton found a natural generalisation of the ordinary complex numbers, having 4 terms (quaternions), for which all laws of algebra are valid except one, the commutative law of multiplication: ab differs from ba. This was the beginning of modern algebra, which can be rightly compared with Riemannian geometry in its influence not only on mathematics but on physics as well. The quaternions themselves have not been as fertile as Hamilton and some of his enthusiastic pupils had hoped. A more general construction, Cayley's matrices, have proved to be the versatile tool for countless mathematical and physical investigations. I learned their use as a young student (Rosanes and Minkowski were my teachers in algebra) and I have applied them to different problems of theoretical physics (for instance in the lattice theory of crystals). So I had the good luck to recognise Heisenberg's symbolic multiplication (which I shall discuss presently) as an example of the well-known matrix calculus, and to be the first, as far as I know, who ever wrote down a strange equation like $pq - qp = h/2\pi i$ (published jointly with Jordan), in which the non-commuting symbols are true representatives of physical quantities (coordinate q and momentum p). Almost simultaneously Dirac established non-commutative mechanics in a very general and satisfactory form. To-day the whole of theoretical physics is based on these mathematical methods.

It is often said that it was a metaphysical idea which led Heisenberg to the principle of matrix mechanics, and this statement is used by the believers in the power of pure reason as an example in their favour. Well, if you were to ask Heisenberg, he would strongly oppose this view. As we worked together I think I know what was going on in his mind. At that time we were all convinced that the new mechanics must be based on new concepts having only a loose connection with classical concepts, as expressed in Bohr's postulate of correspondence. Heisenberg felt that quantities which had no direct relation to experiment ought to be eliminated. He wished to found the new mechanics as directly as possible on experience. If this is a 'metaphysical' principle, well, I cannot contradict; I only wish to say that it is exactly the fundamental principle of modern science as a whole, that which distinguishes it from scholasticism and dogmatic systems of philosophy. But if it is taken (as many have taken it) to mean the elimination of all non-observables from theory, it leads to nonsense. For instance, Schroedinger's wave function ψ is such a non-observable quantity, but it was of course later accepted by Heisenberg as a useful concept. He stated not a dogmatic, but a heuristic principle. He found by an act of scientific intuition the spurious conceptions that have to be eliminated. I shall try to describe this.

According to Bohr, the electrons move in orbits around the atomic nucleus similar to those of the planets around the sun. Classical mechanics uses the method of Fourier to describe such quasi-periodical motions; each coordinate is analysed into a sum of harmonic motions, the first of which has a frequency ν_1, the fundamental, the following ones multiples of it, $\nu_2 = 2\nu_1$, $\nu_3 = 3\nu_1, \ldots$, the overtones.

The line spectra of atoms, however, show no trace of these harmonics, $\nu_n = n\nu_1$. Instead they obey a rule discovered by Ritz. All frequencies in a spectrum can be expressed with the help of a set of 'terms' T_1, T_2, $\ldots$ in the form $\nu_{nm} = T_n - T_m$; they form therefore a quadratic array with two indices

$$\begin{pmatrix} 0 & \nu_{12} & \nu_{13} \cdots \\ \nu_{21}, & 0 & \nu_{23} \cdots \\ \cdot & \cdot & \cdot \cdot \cdot \cdot \end{pmatrix}.$$

The experimental spectroscopists used to arrange their measurements of groups of lines, so called multiplets, in such arrays, and it seems strange now that these never suggested the idea of a matrix to a mathematically trained physicist. But it did not happen this way. The actual progress was much more indirect and based on much more evidence. First came Bohr's identification of the terms T_n with the energy values of stationary states by applying Planck's law in the form $E_n = hT_n$. Then followed a long series of considerations deriving quantum formulae from classical ones by a kind of mathematical guessing, guided by correspondence (Bohr, Heisenberg and Kramers, Born). These formulae, well confirmed by experiment, suggested to Heisenberg the idea that these quantum formulae could be expressed by a kind of symbolic multiplication; the clue to this formalism is the remark that from Ritz's rule there follows the additive combination law of frequencies,

$$\nu_{nk} + \nu_{km} = \nu_{nm};$$

hence the multiplicative combination rule for the amplitudes

$$e^{i\nu_{nk}t} \cdot e^{i\nu_{km}t} = e^{i\nu_{nm}t}.$$

This leads at once to Heisenberg's multiplication, which was soon identified with the well-known matrix calculus (Born and Jordan). The first confirmation of the new theory and its generalisations (Dirac's q-numbers, Schroedinger's operators) was found in the identity of the results in simple cases with the formulae previously obtained by correspondence.

It was therefore an essentially inductive line of reasoning which led to the most abstract theory known in physics, where observables are represented by non-commuting quantities (matrices or operators) and their numerical values by their latent roots or eigenvalues.

Quite different was the origin of wave mechanics. The corpuscular character of cathode rays seemed to be finally settled by J. J. Thomson's experiments and nobody expected that they could be made to produce interference fringes. De Broglie's association of waves with corpuscles is therefore rightly considered as a triumph of intuition. But here also the empirical foundations are quite clear: Special relativity had shown that

(1) the three components of momentum p and the energy ϵ form a 4-vector, i.e. have certain definite transformation properties;

(2) the three components of a wave vector (a vector in the direction of the wave normal of length $k = 1/\lambda$, where λ is the wave-length) and the frequency ν also behave like a 4-vector;

(3) Planck's quantum theory, meanwhile well established by experiment, states that with every energy ϵ there is associated a frequency ν such that $\epsilon = h\nu$.

These facts strongly suggest that with any particle there is associated a wave whose wave vector is parallel to the momentum of the particle and satisfies $p = hk$.

This is de Broglie's law. He studied the consequences for plane waves and indicated the interpretation of Bohr's quantum conditions with the help of standing waves. But what did he predict? As far as I know, nothing. Then were the interference fringes of cathode rays discovered experimentally? There is no truth in this either. The real facts are these: Directed by a remark of Einstein, my colleague Franck and I pondered about the meaning of de Broglie's waves. One day I received a letter from Davisson in America, containing accounts of measurements on the reflexion of electrons by nickel crystals with the question whether we could make sense of the strange maxima and minima of his curves. How it came that we connected these with de Broglie I cannot remember in detail. Some remarks by Einstein had something to do with it; also considerations about the Ramsauer effect (i.e. the increase of the range of free electrons in some gases for decreasing velocity). Anyhow, we encouraged Franck's pupil Elsasser to work it out; he found the correct explanation and de Broglie's formula was confirmed. The final demonstration of electronic diffraction by crystals is due to Davisson and G. P. Thomson. It is a remarkable historical fact that the son of the man who established the corpuscular nature of cathode rays was destined to reveal their undulatory features.

This is the true story, which does not in the least minimise de Broglie's achievement. For such is the natural way of scientific progress.

The real fertility of these ideas was brought to light by Schroedinger. He discovered their connection with Hamilton's work on dynamics and geometrical optics and established the general wave equation which holds not only for free electrons but also for those bound in atoms. To him is due the representation of physical quantities by linear operators. This method has the considerable advantage of transforming the strange equations of non-commutative physics into ordinary analysis. If for instance the momentum p is represented by the differential operator $\dfrac{h}{2\pi i}\dfrac{d}{dq}$ the commutation rule $pq - qp = \dfrac{h}{2\pi i}$ is reduced to the trivial statement that for any function $f(q)$

$$\frac{d}{dq}\,qf(q) - q\,\frac{d}{dq}f(q) = f(q).$$

By substituting such differential operators into the energy equation of mechanics one obtains Schroedinger's wave equation. It must be remembered that this powerful synthesis of wave theory and mechanics sprang into being quite independently of matrix mechanics; the equivalence of the two methods was later proved by Schroedinger, together with numerous applications to old and new problems.

However, the new theory was very formal. Nobody knew what Schroedinger's wave function really meant. Again the solution of this question was no free invention of the mind but forced by experimental facts. The statistical interpretation of de Broglie's waves was suggested to me by my knowledge of experiments on atomic collisions which I had learnt from my experimental colleague James Franck. The whole development of quantum mechanics shows how the accumulation of observations and measurements slowly produces abstract formulae for their condensed description and that the understanding of their meaning follows afterwards. This was accomplished by Heisenberg's considerations on the impossibility of simultaneous accurate measurements of position and velocity, and other such pairs of 'conjugate' quantities, (uncertainty relations) followed by a great number of abstract mathematical investigations bordering on epistemology and philosophy (Jordan, Dirac, Neumann and others).

The essence of the statistical interpretation is this: the square of Schroedinger's ψ-function for a set of particles represents the probability of finding the particles at the places (or with the velocities, or with the energies) indicated by its arguments. It would be very attractive to expand on this fascinating subject, especially on the uncertainty relations and on the question of causality and determinism in physics. But this would be outside the frame of this lecture, and I must restrict myself to a few remarks. According to classical mechanics the configurations and velocities of all parts of a closed system at a given moment completely determine its future motion. In quantum mechanics there is also a quantity which is determined by its initial value, namely the ψ-function; however, one cannot find the configuration and velocities of the particles from ψ, but only the probability of a certain configuration or a certain set of velocities. Therefore the situation with respect to determinism is fundamentally different.

Statistical methods had been used in thermodynamics long before the development of quantum mechanics. They were regarded as an expression of our lack of knowledge, with the idea in the background that this shortcoming might yet be amended. In the new theory there is a natural limit for the improvement of our information, and statistics becomes an integral part of mechanics itself.

Thermodynamical statistics has become a central part of physics, and it is necessary to cast a glance at its development.

Thermodynamics is the classical example of the inductive method. The two fundamental laws, concerning the conservation of energy and the existence and monotonous increase of entropy, are condensed expressions of accumulated experience, namely of the impossibility of constructing a perpetuum mobile, and of a machine which could pump heat out of a reservoir (like the sea) and transform it completely into mechanical work (perpetuum mobile of the second kind). Whittaker has called a statement of this kind a 'principle of impotence' and he has expressed the idea that a few of these principles are sufficient to derive the whole of physics. Relativity, for instance, is the consequence of our impotence to send signals with unlimited velocity, and quantum mechanics can be reduced to our impotence to measure simultaneously coordinates and momenta, etc. However this may be, I wish to stress the point that all these principles, in particular those of thermodynamics, are not *a priori* given, but are the results of a long experience. Man has never acknowledged defeat except after a stubborn struggle. However, in this connection the case of Robert Mayer needs special consideration. He was a physician, and his scientific imagination was directed towards the problem of the equivalence of heat and mechanical work by a physiological observation, the difference of colour of the human blood in the tropics and in our moderate climate. From this strange starting-point he found in the end a method of calculating the mechanical equivalent of heat from simple properties of gases. But when he submitted his paper for publication it was rejected by the editors because it was richly adorned with philosophical and metaphysical considerations. At that time this was no recommendation for a physical theory. Joule's painstaking measurements and Helmholtz's mathematical reasoning on the other hand were accepted without difficulty. This was hard for Mayer, and perhaps not just; for he had given convincing evidence, as was later acknowledged by Joule and Helmholtz. Let us try to learn from this regrettable affair. If we reject some philosophical arguments, that does not mean that we reject any theory to which they are applied. I beg you to remember this when I have to criticise some modern authors.

The amazing feature of thermodynamics is that a few simple and negative statements lead to such far-reaching consequences as the existence of absolute temperature and of entropy, and to a great number of numerical relations between measurable quantities, such as specific heat, compressibility, thermal expansion, galvano- and thermo-electric coefficients, chemical affinities, etc. However, thermodynamics is, in spite of its name, only a formal connection between thermal and dynamical properties. The real identity of heat with motion was established by the kinetic theory, first of the gases, later of systems of a more general kind. You all know the fundamental idea: it is neither possible nor necessary to know each detail of the motion of all the innumerable atoms in a piece of matter, but it suffices to know their average behaviour in order to predict the measurable phenomena. In this way statistics is introduced into mechanics. The principles of statistical mechanics have developed step by step, by trial and error, from the first establishment of Maxwell's distribution law of velocities to the most complex generalisations of Boltzmann, Gibbs, Fowler and Darwin. These principles involve of course the concept of probability and share its controversial character. As far as I see the only foundation of the doctrine of probability, which (though not satisfactory for a mind devoted to the 'absolute') seems at least not more mysterious than science as a whole, is the empirical attitude: The laws of probability are valid just as any other physical law in virtue of the agreement of their consequences with experience. The development of statistical physics is a demonstration of this view. Each statistic depends on the choice of equally probable cases, or, more generally, on the choice of the weight of a given distribution. It is true that the invariance properties of the equations of classical mechanics restrict this choice to some degree (by the so-called theorem of Liouville), but the result that the statistical weight is proportional to the extension in phase-space (coordinates and momenta) can be justified only by the agreement of the consequences with observations.

The same holds for the modifications introduced by quantum theory. The description of the statistical weights is even simpler for quantised systems: each state of given energy which by no physical means can be split into several states has the same weight. This assumption has been checked by numerous examples; if it is for instance applied to the case of electric oscillators emitting and absorbing radiation one obtains for the latter Planck's law.

But just this example can also be considered from a different standpoint and leads then to a new and fundamental result. According to de Broglie the radiation itself must obviously be equivalent to a gas of quanta of light or of photons, to which quantum statistics can be applied directly (without using absorbing and emitting oscillators). If now these photons are treated as genuine particles, having an individuality of their own, Planck's law would not be obtained. One has instead to assume that two states which differ only by the exchange of two photons are physically indistinguishable and have statistically to be counted only as one state. In other words, photons have no individuality. Bose and Einstein have extended this assumption to other gases and shown that for extremely low temperatures and high pressures there ought to be deviations from the ordinary gas laws.

Unfortunately these conditions are hardly attainable by experiment, and the interesting result about the lack of individuality of particles would have remained a theoretical speculation if it had not been endorsed by a quite different way of reasoning.

This came from spectroscopy. The first step was the discovery of the spin of the electron by Goudsmit and Uhlenbeck, which happened before the days of wave mechanics, and was an interpretation of empirical facts in terms of mechanics of the electron. These facts consisted in the observation that many spectral lines showed a fine structure (doublets, triplets, etc.) which could not be explained by assuming the electron as a structureless particle; they could be explained by ascribing a spinning motion to the electron, if this spin was treated by quantum rules already known (Stern-Gerlach effect).

The second step was also connected with spectroscopy. The clue to the understanding of atomic spectra is Ritz's combination principle which we have already discussed (p. 19): all lines of a spectrum can be obtained by taking differences $\nu_{nm} = T_n - T_m$ of a set of terms $T_1, T_2, \ldots$. But it was evident from the beginning that not all of these differences correspond to real observable lines, and so-called selection and exclusion rules had to be formulated. When Bohr had succeeded in interpreting the terms T_n as the energy levels of electronic orbits, and was able to ascribe definite quantum numbers to each of the electrons, it turned out that not only had certain transitions between two electronic states to be excluded, but that even some mechanically possible states did not occur. Pauli formulated this exclusion principle in an extremely simple manner: States in which two electrons would have the same set of quantum numbers (those of the spin included) do not exist; and, moreover, if two sets of quantum numbers differ only by exchanging those of one electron with those of another they represent only one and the same state of the whole atom.

Here again we recognise the lack of individuality of particles, but on the ground of much more direct evidence. For we must not forget that Pauli's exclusion principle rests on such facts of observation as the non-existence of the lowest state of the helium atom (both electrons having the lowest set of quantum numbers) and is supported by innumerable consequences. The most important of these is Bohr's explanation of the periodic system of the elements, which rests essentially on the idea of completed shells of electrons implied by Pauli's principle.

From the standpoint of wave mechanics the situation can be described in this way: Consider a function $\psi(n_1, n_2)$, where n_1, n_2 are the quantum numbers of two indistinguishable particles. As mentioned before the square of ψ represents the probability of finding the particles in the states n_1, n_2, and the lack of individuality is expressed by $\psi^2(n_1, n_2) = \psi^2(n_2, n_1)$. There follow two possibilities for ψ itself,

either $\qquad \psi(n_1, n_2) = \psi(n_2, n_1),$

or $\qquad \psi(n_1, n_2) = -\psi(n_2, n_1).$

The second case implies for equal values of n_1 and n_2

$$\psi\,(n,\,n) = -\,\psi\,(n,\,n) = 0,$$

which is obviously the expression of Pauli's exclusion principle. Now it has turned out that not this case, but the first one (ψ symmetric in its arguments), corresponds to the statistics of Bose and Einstein, while the other case (ψ skew in its arguments) indicates a quite different behaviour. The statistical consequences for this case, which holds not only for electrons, but also for protons (and other particles), have been worked out by Fermi and Dirac.

These symmetry properties of the wave functions and Pauli's principle are an essential part of quantum mechanics. I hope to have convinced you that they are derived by a long inductive process, in which flashes of imagination have alternated with painful observation and interpretation of facts. It was a period of ideal co-operation of experiment and theory. There was neither the experimenter boasting about the empirical purity of his results, nor the theorist claiming *a priori* knowledge, but mutual help and encouragement.

Once established, quantum mechanics and quantum statistics of course permitted countless analytical predictions to be made, many of which were confirmed by experiments. The electronic structure of atoms and molecules could be subjected to calculations similarly to the planetary system in the century after Newton.

The most important results are the explanation of line and band spectra, of the nature of the metallic state and of chemical valency. The number of predicted or confirmed experiments was overwhelming. One of the most admirable predictions was that of the existence of two types of hydrogen atoms, para- and ortho-hydrogen, by Heisenberg.

Theoretical physics seemed to be definitely and triumphantly ahead of experiment.

But, alas, only for a short time. Again there came a wave of experimental discoveries, many of which were a complete surprise and not even implicitly contained in the accepted theory.

Most of these are concerned with radioactivity, cosmic rays and the atomic nucleus. Perhaps the most unexpected discovery was that of the neutron (Chadwick). It completely changed our ideas about the structure of the nucleus and opened the way to treatment by quantum mechanics. In general it can be said that from about 1930 experiment was ahead of theory. But there are two outstanding theoretical predictions, namely of new elementary particles, the positron and the meson.

However admirable, these discoveries are not products of pure reason, but the final outcome of a long chain of empirical research. The positron is connected with Dirac's linear wave equation of the spinning electron. To give you the history of this equation would need a whole lecture. It may suffice to say that the spin of the electron, which, as explained above, was discovered by typical induction from spectroscopic facts, the multiplet nature of certain lines (Uhlenbeck and Goudsmit), was fitted into quantum mechanics by the introduction of a set of simple matrices representing the two directional states of the spin (Pauli), and that every step of the development of the spin theory was inspired and checked by spectroscopic evidence. Dirac gave it the last polish by discovering that the natural relativistic generalisation of Schroedinger's wave equation led automatically to the spin. How the discussion of the solutions of Dirac's equation brought to light states of negative energy, and how Dirac succeeded in reconciling these with our traditional ideas about energy by the interpretation of non-occupied states as positrons is too technical to be recorded here. But I must mention that he first believed that the positive particles predicted in his theory were protons, and was corrected by experience; he recognised them as positive electrons when these were actually discovered in cosmic rays (Anderson, Blackett).

In the case of the meson, I shall try to indicate some of the ideas which led the Japanese physicist Yukawa to the suggestion of the existence of new particles with a mass midway between electron and proton. The starting-point was the existence of forces of very short range which keep the neutrons and protons in the nucleus together. Yukawa observed that a potential of the form $e^{-r/a}/r$ would have the properties required, if the constant length a were chosen of the order of nuclear dimensions (10^{-13} cm.); this potential is a generalisation of the Coulomb potential $1/r$ of the electrostatic forces (charge 1) and satisfies, not the Laplace equation $\Delta\phi = 0$, but the slightly modified equation $\Delta\phi = \phi/a^2$. Now just as electrostatics can be considered as a special case of electrodynamics, described by Maxwell's equations, one can construct a dynamical Yukawa field which contains the one given above as the statical case. Each field component satisfies, instead of the ordinary wave equation $\varDelta\phi - \frac{1}{c^2} \frac{\partial^2 \phi}{\partial t^2} = 0$, Yukawa's modified equation

$$\varDelta\phi - \frac{1}{c^2} \frac{\partial^2 \phi}{\partial t^2} = \frac{\phi}{a^2}. \tag{4}$$

The complete field equations can be obtained from an action principle which differs from that of electrodynamics given above ((2), p. 7), by a term $\frac{1}{2}a^2 (\phi^2 - A^2)$ added to the integrand.

Now these equations have solutions which represent plane waves, just as in Maxwell's case, and according to de Broglie's general principle there must be particles associated with these waves in the same way as photons are associated with light waves. According to the theory of relativity the energy ϵ of a particle of mass m depends on its momentum p by the formula

$$\left(\frac{\epsilon}{c}\right)^2 = m^2 c^2 + p^2. \tag{5}$$

If one introduces here Planck's and de Broglie's relations $\epsilon = h\nu$, $p = hk$, one obtains

$$\left(\frac{\nu}{c}\right)^2 = \left(\frac{mc}{h}\right)^2 + k^2. \tag{6}$$

On the other hand Yukawa's wave equation (4) leads for a wave $\phi = A \sin 2\pi (\nu t - kx)$ to the same relation between ν and k, provided

$$\frac{mc}{h} = \frac{1}{2\pi a}. \tag{7}$$

This is a relation between the range a ($\sim 10^{-13}$ cm.) of the nuclear forces and the mass m of the particles

33

associated with the vibrating Yukawa field, and one obtains

$$m = \frac{h}{2\pi ac} = \frac{6 \times 10^{-27}}{2\pi \times 10^{-13} \times 3 \times 10^{10}} = 3 \times 10^{-25} \text{ g.}; \quad (8)$$

this is some hundred times bigger than the mass of the electron (10^{-27} g.), but definitely smaller than that of the proton ($1800 \times$ electron mass). In this way, Yukawa was led to predict a new particle, now called meson, which was actually discovered a short time later in the cosmic radiation.

Again this does not appear as a result of *a priori* principles but an ingenious synthesis of well-established knowledge with a new simple assumption.

These examples I hope will suffice to show you the way theories are formed and used.

Charles Darwin, my predecessor in my Edinburgh chair, once said something like this: "The Ordinary Man can see a thing an inch in front of his nose; a few can see things 2 inches distant; if anyone can see it at 3 inches, he is a man of genius." I have tried to describe to you some of the acts of these 2- or 3-inch men. My admiration of them is not diminished by the consciousness of the fact that they were guided by the experience of the whole human race to the right place into which to poke their noses. I have also not endeavoured to analyse the idea of beauty or perfection or simplicity of a natural law which has often guided the correct divination. I am convinced that such an analysis would lead to nothing; for these ideas are themselves subject to development. We learn something new from every new case, and I am not inclined

34

to accept final theories about invariable laws of the human mind.

But now I have to return to my starting-point and to apply the results obtained to the present unsolved problems of physics, and in particular to Eddington's philosophy. In spite of the brilliant achievement of the last period the state of theoretical physics is just as problematical as it was at any time—if we except the proud late-Victorian days when all riddles were believed to be solved. There are several elementary particles: photons, electrons and positrons, neutrons and protons, mesons charged and uncharged (neutrettos), and perhaps neutrinos. Each of them is associated with a wave function and has characteristic constants (mass, charge) which appear in the wave equation. But all these fields are hardly related to another, and we have no theory explaining the dimensionless ratios of the different constants (for instance the ratio 1845 of proton mass to electron mass). Particularly mysterious is a dimensionless number which because of its first appearance in spectroscopy (Sommerfeld) is called the fine structure constant, the ratio $hc/2\pi e^2$ (e charge of the electron) which is empirically near to 137. We have a quantum theory of the nucleus, which has yielded many important results; but it looks just as provisional as Bohr's quantum theory before the discovery of wave mechanics. Much more serious is the disease of the 'infinities'. They are of two kinds, represented by two simple cases: The electrostatic energy of a charged sphere of radius r is, apart from a numerical factor, e^2/r; it becomes infinite for vanishing radius. Hence a point charge has an infinite energy (or mass, according to Einstein's law). Further, the energy of a quantised oscillator is not $h\nu n$ (as Planck originally assumed), where n is an integer, but $h\nu \, (n + \frac{1}{2})$; hence there exists (for $n = 0$) a zero-point energy $\frac{1}{2}h\nu$, and each system which can be considered (by Fourier's theorem) as the superposition of an infinite number of harmonic oscillators (e.g. a cavity filled with radiation) has therefore an infinite zero-point energy. Similar infinities appear in many considerations concerning the interaction of particles and radiation, and a great amount of ingenuity has been expended on getting rid of them. Dirac, Pryce and others have modified the definitions of energy and momentum in mechanics and electrodynamics in order to eliminate the infinite self-energy. I myself have concluded that this infinity shows a fundamental deficiency of Maxwell's equations and have replaced them by a generalised set, and although these equations are non-linear and apparently untractable, they have some chance of being nearer the truth as they appear in a new theory of Schroedinger, which is a synthesis of gravitation, electrodynamics and meson theory. The other type of infinities produced by the infinite number of frequencies has also been tackled with more or less success (by Dirac, Heitler and Peng, and others).

Eddington has been worried by all these difficulties just as much as all of us, and he has made interesting attempts to overcome them. His leading idea is that an essential difficulty in a theory can always be traced to an epistemological error, to a wrong or too narrow concept. With that I agree. But when he tries to correct these errors by constructions on what he considers as epistemological evidence I am reluctant to follow. I am doubtful about the visibility in front of his nose. For instance, he obtains numerical values for two of the dimensionless numbers mentioned above, namely $hc/2\pi e^2$, and the ratio proton mass to electron mass.

They are expressed in terms of the properties of an abstract 'phase space'. The mathematical theory of this space of E-numbers is very beautiful and presented in a way which suggests the most natural construction of the mind. But the fact cannot be denied that nobody had ever considered the E-numbers before Dirac's theory of the spin was developed, which itself is the end of a long series of abstractions forced on us by experiment. And if the E-numbers should have existed in mathematics, as several similar non-commutative 'algebras' have, nobody would have divined that they could be used in physics nor what meaning they might have, before the existence and the properties of the spin had been extracted from observations. Eddington connects the dimensionless physical constants with the number n of the dimensions of his E-spaces, and his theory leads to the function $f(n) = \frac{1}{2}n^2(n^2+1)$ which, for consecutive even numbers $n = 2, 4, 6, \ldots$, assumes the values 10, 136, 666 ....*

* Apocalyptic numbers, indeed. It has been proposed that certain well-known lines of St John's revelation ought to be written in this way:

And I saw a beast coming up out of the sea having $f(2)$ horns... and his number is $f(6)$....

But whether the figure x in

...and there was given to him authority to continue x months...

is to be interpreted as $1.f(3) - 3.f(1)$ or as $\frac{1}{3}[f(4) - f(2)]$ can be disputed.

Now at the time when Eddington began his work the experimental value of $hc/2\pi e^2$ was near to $136 = f(4)$. Later experiments indicated a larger value, and to-day it is very near to 137. Accordingly Eddington adapted his theory by adding a unit.

The mass ratio was also obtained in terms of these integers, namely as the ratio of the two roots of the quadratic equation

$$10x^2 - 136x + 1 = 0,$$

which is $1847 \cdot 9$, near to the experimental value $1836 \cdot 5$ (and has been further corrected).

I cannot criticise the derivation of these expressions as I have not succeeded in understanding them. Anyhow a few coincidences of this kind, which are not true predictions, but expressions of experimentally known quantities, seem to me only a weak evidence for a great theory. And there are hardly any other predictions. Neither the neutron nor the meson has been forecast. And if the number of particles in the universe is calculated there is little hope of checking it experimentally— although I admit that it may be a very useful concept. I am far from attacking Eddington's theories or from doubting his results. If they should turn out to be right I shall rejoice. But I shall not attribute this (possible) success to Eddington's philosophy, as a doctrine which could be followed by others, but to his personal genius and intuition.

Let us now finally have a look at the theory of Milne, which I mentioned at the beginning. He also claims to have derived universal laws of nature by pure epistemological principles. One of these is the 'operational method' of definition. This name has been given by the American physicist Bridgman to a procedure quite common amongst scientists. It consists in the demand that a physical quantity must not be defined by verbal reduction to other familiar conceptions, but by prescribing the operations necessary to produce and to measure it. This is a sound rule, a reaction against verbalism and word fetishism. It is very useful in classical physics where one has to do with quantities accessible to direct measurements, as in thermodynamics (Bridgman himself is an expert in this field) or in electrodynamics. For instance, it is reasonable to introduce temperature by describing the thermometric operations, or to define the electric field by referring to the forces on small charged test bodies. But the operational definition is rather out of place if you wish to extend the idea of the field to atomic nuclei and electrons and it comes to grief in quantum theory. Wave mechanics has a catalogue of 'observables'. But that does not mean that the corresponding quantities are represented by variables whose values could be measured; they are represented by differential or integral operators whose eigenvalues can be measured. I cannot see what experimental 'operation' could be devised in order to define a mathematical operator. Moreover I have already mentioned that there are concepts used in wave mechanics which are not observables, for instance, Schroedinger's wave function; there are in principle no means to observe it, hence no 'operational' definition. However, this may not matter for Milne, as he is mainly interested in astronomy. Let us see how he uses the operational principle.

All empirical knowledge about the stars is based on

the light coming from the sky and its interpretation with the help of terrestrial instruments, telescopes, spectroscopes, clocks. Milne restricts his fundamental consideration to the use of clocks alone. He refuses to admit distance in intrastellar space as an observable and proposes to reduce this conception to statements about the arrival of time signals. For this purpose he supposes the existence of observers on other celestial bodies who have clocks and time signals available like ourselves. The elementary operation for investigating space consists in this: we send out a light signal at a given time t_1 on our clock; this reaches an observer on another star and is reflected or returned back to us, where it arrives at the time t_2 of our clock, carrying with it information about the time τ of arrival of the signal read on the clocks of the distant star. From these data t_1, t_2, τ (frequently repeated) Milne proposes to derive the foundations of geometry, kinematics and even dynamics of the Universe.

Milne's clocks and light signals are of course an imitation of the well-known method used by Einstein in order to show that the idea of absolute simultaneity is absurd, and to derive the Lorentz transformation. But there is a fundamental difference: Einstein's light signal travels only between two stations on one and the same celestial body, not from one star to the other. His model is only an abstraction from and simplification of an actual observation, expressed by the negative result of Michelson's and Morley's experiment. Besides, the Lorentz transformations were not found in this way (they would then be called Einstein transformations), but were already known, derived (by Lorentz)

40

from a study of Maxwell's equations and used for the interpretation of Michelson's experiment. As usual in physics, Einstein's epistemological derivation came after the formal discovery, as an eye-opener for those who stuck to the idea of an absolute time.

Milne's operations on the other hand are no idealisations of real experiments. They seem to me weird inventions: Terrestrial light reflected from a fixed star, whoever has seen such a thing? Or is it likely that in the whole future history of mankind anybody will see it? And observers with clocks on other stars, clocks illuminated by the light coming from us so that we can read them on their return—or observers giving us politely the time on their clocks whenever our light signal arrives. I am at a loss to recognise in such phantasies the realistic idea of operational definition.

But that is not all. Milne has other assumptions and epistemological principles. His observers are not on any arbitrary star, but on the spiral nebulae or galaxies which according to modern stellar astronomy are island universes, innumerable and fairly equally distributed through space. Each of them consists of millions and millions of stars, just as our own galaxy. But Milne considers them as single particles, points with a mass attached to them. Then he uses a principle of uniformity, namely that an observer will find the same laws of nature and the same general aspect of the universe (with the help of his clocks and time signals described above) on whichever galaxy he is situated. This is called an epistemological principle. But one empirical 'fact' is also used, namely, in Milne's formulation, the existence of an absolute zero of time, an epoch of creation. Behind this strange assumption is a set of astronomical observations, indeed, namely the red shift of the spectral lines of the spiral nebulae. It has been found that this red shift increases with increasing distance of the nebulae, and if it is interpreted as a Doppler effect, it indicates that the whole system of galaxies is expanding in all directions. Much speculation has been provoked by this strange fact, and it has been shown (by Lemaître, Robertson and others) that Einstein's general theory of relativity allows solutions of the field equation which represent such an expanding universe. But as only the instantaneous distribution of velocities is observable there remains a wide field of hypothesis about the past and future of the world, whether the Universe is vibrating as a whole, or expanding from an initial state of tremendous concentration. Milne chooses the latter alternative and states it as a fact: The world was created 2×10^9 years ago as an accumulation of masses in a small space.

Milne's explanation of the expansion process which has taken place since this epoch consists in the simple remark that the masses crowded together in the beginning of time had different velocities and were therefore in every moment arranged in such a way that the higher the velocity of a star the more remote it would be. This is indeed the result of the observations of the red shift of spectral lines when interpreted as Doppler effect. Schroedinger once remarked to me that the situation reminds him of a man looking at a horse race in progress and wondering why the horses in front are apparently the fastest, when suddenly he is deeply impressed by the obvious 'kinematical' interpretation.

The same seems to have happened to Milne in such a degree that he has devoted an incredible amount of ingenuity to reconciling this assumption with the principle of relativity and to deriving from it, together with his postulate of uniformity, the laws of Nature, including gravitation and electro-magnetism.

An ordinary physicist can hardly follow this high flight of ideas. Interstellar space is not empty, and matter reacts with light (by dispersion, absorption and re-emission). Light signals travelling over the tremendous distances between galaxies might be essentially affected by such interaction. It seems not sound to exclude this possibility from the beginning by founding the whole of geometry and kinematics on light signals, quite apart from the other objections made above. In fact if one admits Milne's method of surveying space with light signals and clocks only between relatively near neighbour systems (where interstellar matter could be neglected) one obtains, as Robertson has shown, an Einstein universe of the type already mentioned (Lemaître, Robertson). It must further be considered that the interaction of light waves with interstellar matter may contribute to, or even produce, the red shift.

Finally, the age of Milne's universe seems to be rather short compared with that of our planet. According to reliable radioactive methods the rocks of the earth's crust have been formed at least $1 \cdot 5 \times 10^9$ years ago, i.e. three-quarters of the duration of the whole universe.

But I am no expert in these cosmological questions, and I do not intend to stress some weak points of the theory. That would be unjust, for the audacity of the

43

idea of deducing the structure of the world from a few principles, and the skill in overcoming formidable difficulties, can but be admired. I do not wish to discourage anybody who feels in himself the vocation to embark on so adventurous a journey.

But I believe that there is no philosophical highroad in science, with epistemological signposts. No, we are in a jungle and find our way by trial and error, building our road *behind* us as we proceed. We do not *find* signposts at crossroads, but our own scouts *erect* them, to help the rest. Eddington's and Milne's ideas may be such signposts. The difficulty is that they point in opposite directions: Two theories both claiming to be built on *a priori* principles, but widely different and contradictory.

My lecture will have attained its purpose if you find in this contradiction nothing to be wondered at; it is exactly what the empirical scientist would expect. My advice to those who wish to learn the art of scientific prophecy is not to rely on abstract reason, but to decipher the secret language of Nature from Nature's documents, the facts of experience.

For EU product safety concerns, contact us at Calle de José Abascal, 56–1°,
28003 Madrid, Spain or eugpsr@cambridge.org.

www.ingramcontent.com/pod-product-compliance
Ingram Content Group UK Ltd.
Pitfield, Milton Keynes, MK11 3LW, UK
UKHW010851090126
466816UK00011B/160